ISBN 978-1-332-81950-8
PIBN 10469331

Technical Paper 66. (Petroleum Technology 14)

DEPARTMENT OF THE INTERIOR
BUREAU OF MINES
JOSEPH A. HOLMES, DIRECTOR

MUD-LADEN FLUID APPLIED TO WELL DRILLING

BY

J. A. POLLARD AND A. G. HEGGEM

First edition. December, 1913.

CONTENTS.

ILLUSTRATIONS.

MUD-LADEN FLUID APPLIED TO WELL DRILLING.

By J. A. POLLARD and A. G. HEGGEM.

INTRODUCTION.

The Bureau of Mines is investigating the technology of petroleum and its products, the investigation including well-drilling methods and the transportation, treatment, and utilization of petroleum and natural gas, with especial reference to prevention of waste and increased efficiency in the development of oil and gas resources in land belonging to or controlled by the Government.

One of the greatest wastes of natural gas is that which often takes place in drilling oil wells. If a well is being drilled by one of the usual methods, the gas becomes a hindrance to drilling, and the driller regards it as a nuisance; or the gas may be found in a field where it has little or no immediate commercial value, and hence is allowed to escape into the air without restraint. For preventing this waste the usual dry-hole methods of drilling are unsuitable, and it is the purpose of this paper to outline a method whereby wells may be drilled and the oil recovered without waste of gas.

When an open hole is bored into a bed containing gas under pressure, the gas flows toward the hole because of the reduction of pressure at the hole. The movement of gas is therefore always from a greater to a lesser pressure. If some means be provided for keeping the pressure within the well greater than the pressure in the gas sand there will be no flow of gas into the well. The requisite pressure may be obtained by a column of water in the well, provided the gas pressure be not greater than that of the water when the well is full. However, the use of clear water is sometimes impracticable and is always undesirable. The action of clear water on the walls of the well causes caving, and an attempt to use clear water in drilling the well invites trouble and may injure the producing sands. By mixing clay with the water the results obtained are entirely different, and this paper deals with the use of a mud-laden fluid as applied to the dry-hole method of drilling with a cable rig, one type of which is shown in figure 1.

The use of clay-laden water, while not new in well drilling, having been used with rotary rigs for years and employed in 1901 for drilling the first successful oil well in the Beaumont (Tex.) field, was not applied previous to 1913 in drilling by the dry-hole method with a

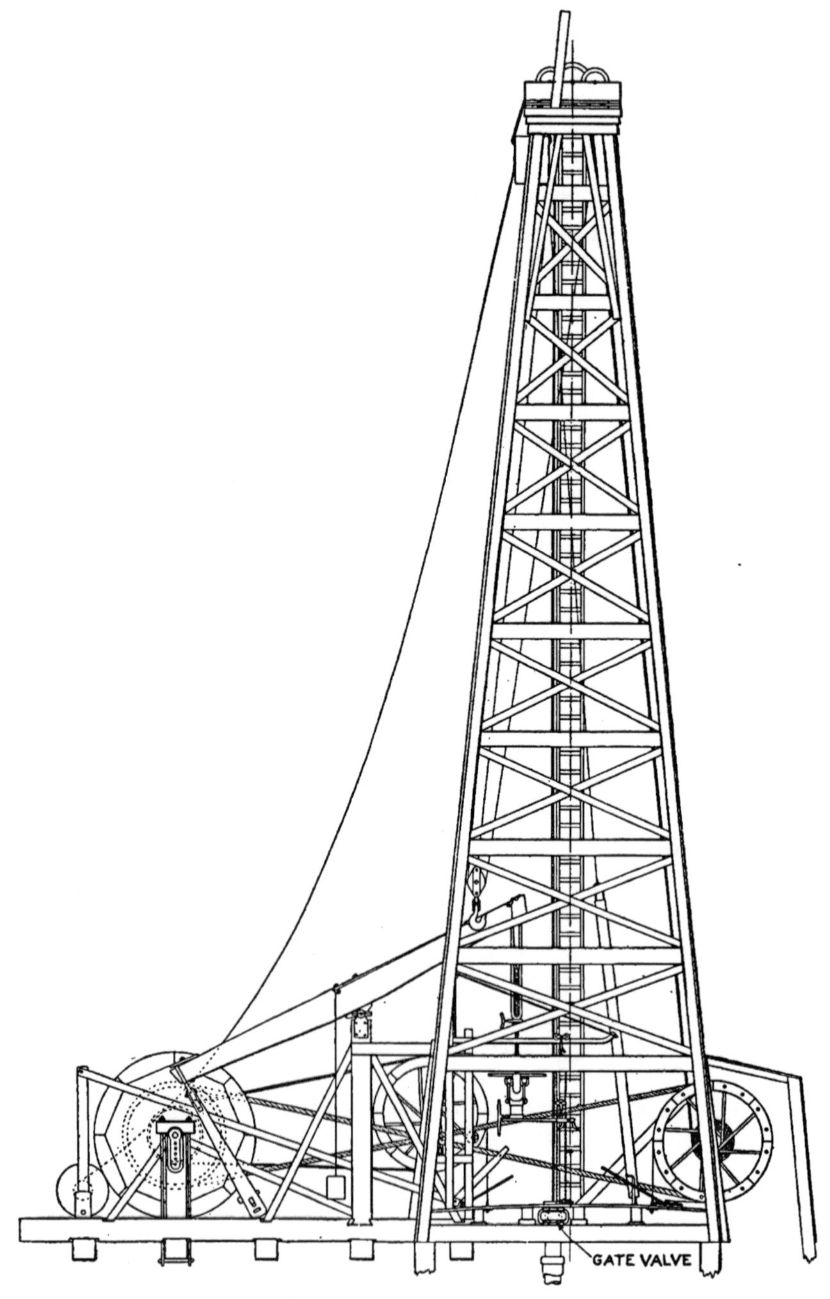

FIGURE 1.—An Oklahoma cable rig, with calf wheel.

cable rig. Already the advantages of the method have been demonstrated, and there can be no question as to its efficiency when properly used. Too much emphasis can not be placed upon the importance of using it where gas and water are encountered, for it not only greatly reduces the danger to workmen, but effects a great saving in the amount of casing needed and entirely eliminates the waste of gas while drilling is in progress.

In the Mid-Continent field alone during the past year there have been a large number of deaths and serious accidents from blow-outs of gas from wells being drilled. These accidents and the great hazard from fire risk, to say nothing of the great waste of gas, can not happen if the mud-laden fluid method be properly used.

In one small field as much as 100,000,000 to 150,000,000 cubic feet of gas a day has been wasted in the effort to obtain oil. Such great waste is believed to be altogether unnecessary, for the preventive methods that are described in this paper have been shown to be entirely practicable. Aside from increasing safety and preventing waste, the methods offer a further advantage in that they absolutely prevent the contents of one bed mingling with those of another; thus water can not enter the pay sands, neither can oil or salt water contaminate the fresh water of other beds.

DESCRIPTION OF MUD-LADEN FLUID.

In this paper the term "mud-laden fluid" is applied to a mixture of water with any clay which will remain suspended in water for a considerable time. The fine sticky clays that in many places are termed "gumbo" are well suited for this purpose.

Some oil workers have thought that "mud-laden fluid" implies the use of any of the drillings from the well; but this is not the case, for if any coarse material in the drillings, such as sand, is used it will settle in the well and prevent the bit from striking the bottom of the hole. The proportion of clay that should be mixed with water to insure the best results is about 20 per cent by weight. With this proportion of clay in the water it is impossible for the driller, no matter how experienced he may be, to tell whether there is any clay at all in the hole, for the tools work about the same as they would if the hole were filled with clear water. An excellent idea of the consistency required can be obtained by comparing the action of a stream of sand pumpings, or muddy water, running in a ditch with that of clear water. The sand pumpings contain fine material that is deposited on the walls, and especially the bottom of the ditch, where it forms an ever-thickening protective coating; clear water, on the other hand, cuts away the sides and bottom of the ditch and may cause it to cave. Between clear water and water containing more mud than can be held in suspension by the current, it is possible to find a mixture of clay and water that will deposit part of the clay as a fine, protective coating while the rest of the clay remains in suspension and passes through the ditch.

THE ACTION OF MUD-LADEN FLUID ON POROUS FORMATIONS.

The action of the mud-laden fluid on gas rock or gas sands, or other porous formations, can be likened to the action of muddy water going through a filter. In any filter that has been used for some time, it will be found that most of the sediment from the water has been deposited on the surface of the filter, but some of it has entered the filter, the proportion diminishing with the distance penetrated.

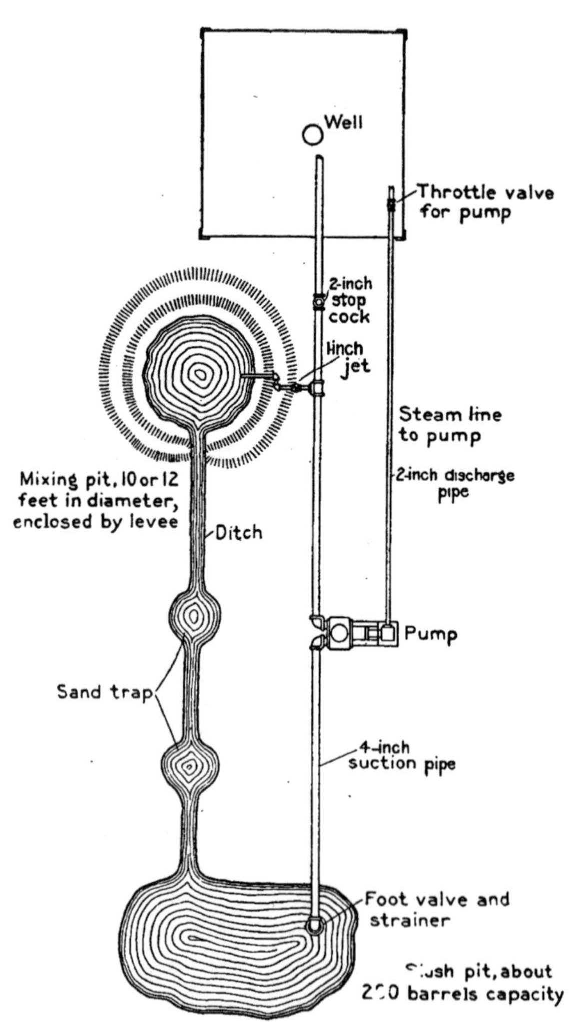

FIGURE 2.—Arrangement for handling mud-laden fluid.

The distance to which clay from the fluid in a well will penetrate a porous formation depends on the excess of pressure produced by the column of fluid or by the pump, and also on the porosity of the formation, but finally no more water will go through.

Some drillers contend that clear water should have the same effect as the mud-laden fluid, but the results of trials have shown that it does not. Many wells can not be filled with clear water, because the water continues to flow into the rock or sand without any clogging effect and in consequence does not rise high enough in the well to give a pressure sufficient to overcome that of the gas. Drillers have attempted this method, using clear water, and have permanently drowned out a gas sand. Further than this, clear water causes the walls of the well to slack and cave and " freeze " the pipe.

The action of the muddy water is entirely dissimilar. The fluid enters the porous stratum for a short distance, and deposits clay that clogs the openings and finally prevents the further inflow of fluid.

REOPENING A SEALED BED.

Should it be desired at any time to recover the gas from a porous bed that has been clogged, all that is necessary is to bail down the fluid in the well until the pressure of the remaining fluid is less than the gas pressure. The fluid will then be forced out of the well by the superior gas pressure and the gas sand will be thoroughly cleaned of all mud. By using this method the operator can seal or unseal at will a gas-bearing formation.

PREPARING THE MUD-LADEN FLUID.

In order to save time in preparing the clay and water mixture for a well it is recommended that a slush pit, about 15 or 20 feet long, 10 feet wide, and 3 feet deep, be dug close to the derrick (see fig. 2). The place for this pit does not matter much, except that it should be on the lowest side of the derrick, so that when the well is bailed the fluid will run into the pit without trouble. When a well is being drilled through beds of clay the drillings from these beds can be turned into the pit as they come from the well and thus be saved and kept from becoming mixed with sand and shale drillings. Care should be taken not to mix with this fluid any material that will not stay in suspension. Not more than half a day's labor is necessary to prepare the fluid for the well, and the work can be done by ordinary unskilled laborers.

The pump recommended for use in handling the mixture of mud and water is known in the oil business as a " duplex slush pump," fitted with removable liners and rubber valves. These pumps may be obtained in many sizes, some of them weighing about 4,000 pounds. Such heavy pumps are costly and are expensive to move from one well to another; consequently the old-style 8 inch by 5 inch by 10 inch pump, which weighs less than 2,000 pounds, seems best adapted for this work.

METHODS OF INTRODUCING FLUID INTO WELLS.

There are several methods of introducing the clay and water into a well.

Before gas is encountered in a well that has been drilled in the most advantageous manner, several hundred feet of the hole may be

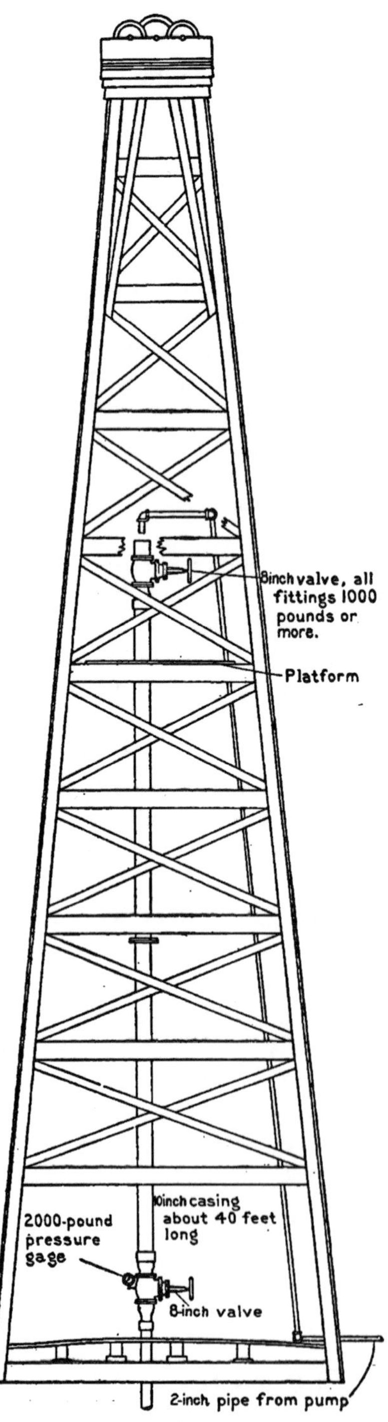

FIGURE 3.—Device for introducing mud-laden fluid into a well under gas pressure.

without casing. To prevent the walls from caving, as might happen were the fluid pumped directly into the top of the well, a string of tubing reaching to the bottom of the well should be placed to conduct the fluid. The fluid is then pumped in until the well is filled.

PUTTING THE FLUID IN A WELL THAT IS BLOWING GAS.

If, after gas has been struck, the well is blowing gas, and the conditions are such that the gas can be shut in, recourse may be had to a method which has been named the "lubricator system," shown in figure 3, which consists preferably of two joints of 10-inch casing placed above a master valve on the head of the well and having a second gate valve at the top. These valves and casings can be most readily attached to the well by assembling them on the ground and placing them on the well as a unit. It is dangerons to attempt to put a valve or a single fitting on a gas well by handling the valve or fitting in slings.

As soon as the valves and the two joints of casing, which are termed the "lubricator," are in place, the bottom valve is closed. The mud-laden fluid is then pumped into the two joints of casing, and when they are entirely filled the upper valve is closed and the bottom valve is opened. Following the equalization of pressure throughout the device, the mud-laden fluid drops to the bottom of the well. As soon as the fluid has passed the lower gate valve, as shown by the sound when the casing is struck, this valve is closed and the upper valve is opened. The volume of gas that escapes from the hole is equal to the volume of fluid that has been introduced, and therefore the pressure of the gas in the casing is not increased. After a few repetitions of this operation a part of the fluid is forced out of the bore hole into the porous strata. Then the gas remaining in the hole will expand and its pressure will be lowered. The amount the pressure is reduced is an indication of the amount of fluid forced into the porous formations. By repeating the operation described there is finally established in the well a column of fluid sufficient to overcome the gas pressure, and then the remaining space can be filled by pumping directly into the casing.

When gas is blowing from a well and can not be shut in, perhaps because of the small amount of surface casing in the well or possibly because of the casing not being properly seated, so that gas is forced up outside the casing when the valve is closed, another method is used.

PUTTING THE FLUID IN A WELL THAT CAN NOT BE SHUT IN.

The method of forcing mud-laden fluid into a gas well that can not be shut in is illustrated in figure 4.

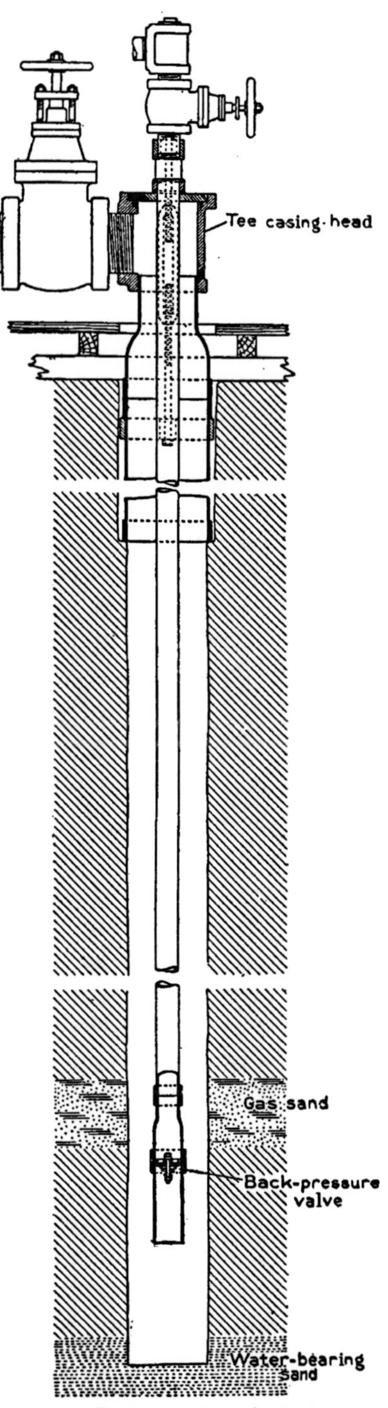

FIGURE 4.—Equipment for introducing fluid into a well that can not be shut in. Well section is generalized.

At such a well it becomes necessary to insert a string of tubing, with a back-pressure valve at the bottom, to a point below the gas sand. After the tubing has been lowered to the proper depth it is packed or sealed off with a casing-head tee previously placed on the well. To control the flow of gas a gate valve should be placed on the lateral discharge of this tee. As soon as the mud-laden fluid is started down the well through the tubing, the gate valve on the tee can be partly closed in such a manner as to throttle the outlet and to prevent the mud-laden fluid from being forced out of the well by the gas pressure.

The amount of throttling necessary can be determined only by the man in charge, as similar conditions will not prevail at any two wells. However, it is not difficult to ascertain how much the gas should be throttled to stop the fluid from being thrown out of the well. If a well is emitting water with the gas, the fluid can be put in just as readily in this manner, and a well with a capacity of 40,000-000 cubic feet of gas and several thousand barrels of water daily can be controlled in 15 or 20 minutes. As is evident, the full rock pressure of the well is not maintained in the casing, and consequently no blow-out follows, as would happen were the gas forced to the surface on the outside of the casing.

PROCEDURE WHEN THE PRESSURE OF THE COLUMN OF FLUID IS INSUFFICIENT.

Sometimes the gas in a sand has a greater pressure than that of the mud-laden fluid in the well. When this happens the fluid is blown out and the well becomes wild. It is then necessary to use a pump to establish a greater pressure in the muddy fluid than that of the gas, in order to force the mud into the sand. A sufficient pump pressure should be allowed to remain on the well for at least two or three hours and then relieved very slowly and carefully. If the pressure is suddenly reduced, not only may the column of fluid be violently ejected and the casings, fittings, and derrick wrecked, but the well may cave. If the extra pressure is maintained for several hours and then released slowly, the tools can be put into the well and drilling resumed, because the porous bed is clogged around the tools and also below them to some depth. Great care should be taken, however, not to drill too deep at one time without applying the necessary pump pressure, because the clogged portion is shallow and may be drilled through in a short time. By repeating the procedure just described, the well can be drilled through the formation in which the gas pressure is greater than that exerted by the column of fluid. Each time that pressure is applied by the pump the mud-laden fluid is driven into the bottom and sides of the bore hole, thus excluding the gas from the path of the bit, so that when drilling starts again the tools are drilling in a formation that has been filled with the clay from the fluid.

DRILLING WITH FLUID IN THE HOLE.

The presence of the mud-laden fluid within the well does not interfere with drilling. The bailer can be used in the usual way to remove the drillings from the bottom. In the usual dry-hole method of drilling through a formation from which gas is escaping the drillings are blown out of the well and can not readily be saved for examination, whereas with the mud-laden fluid method samples of the formation are readily obtained.

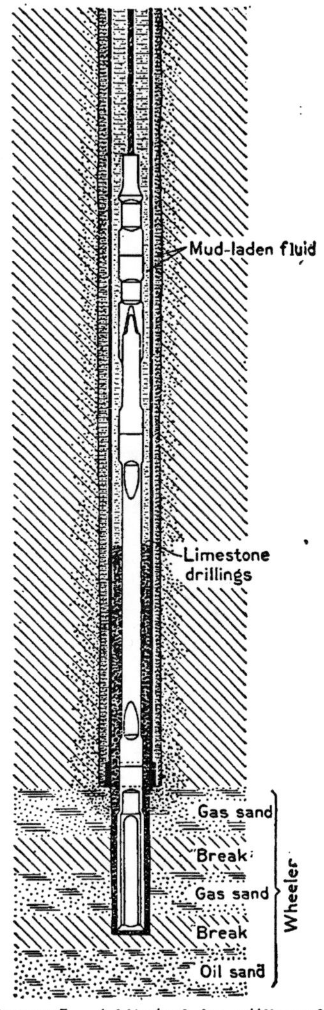

FIGURE 5.—A bit stuck by settling of drillings. Well section is generalized.

It has been stated by some that in hard limestone too much time would be lost if mud-laden fluid were used, because the tools would not drop readily in the fluid. However, at some wells in the Cushing (Okla.) field as much as 22 feet has been drilled through the Wheeler limestone in 18 hours, so that apparently the tools work better in the fluid than they would if the gas were blowing, as it often takes from six to eight days to drill through such a gas sand. Not only is less time consumed by the new method, but the risk of fire and danger to workmen from blowouts is obviated. Many instances can be cited of gas pressures so strong that it was impossible to drill through the gas sand into the oil sand below, and, consequently, the well had to be shut in and called a "gasser," though perhaps there was no market for the gas. In drilling gas wells by the mud-laden fluid method the well can be sealed when gas is struck, thus maintaining the initial rock pressure of the well, so that if it be desired to drill a number of gas wells near each other the rock pressure of the later wells will not be decreased by the drilling of the earlier wells.

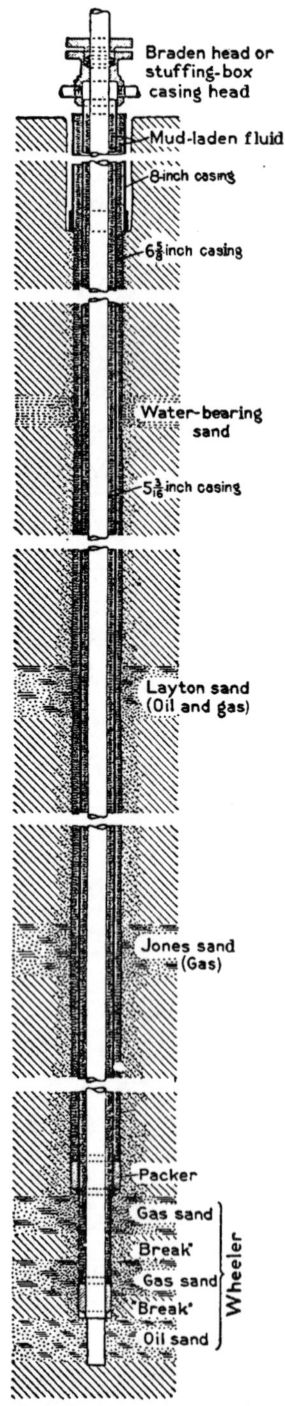

FIGURE 6.—Generalized section of well finished with mud fluid. By removing the fluid from between the casings a "combination" well is made.

In drilling through limestone with the mud-laden fluid in the bore hole great care should be taken not to drill too much hole at one time, as experience has proved that limestone drillings, when too much hole has been made, will settle back around the tools and "freeze" them in the well while the temper-screw clamps are being removed and the bull ropes thrown on preparatory to withdrawing the tools. The "freezing" of the tools is shown in figure 5.

From observations of engineers of the bureau, it is recommended that not more than 3 feet of hole should be drilled at one time in limestone with the fluid in the well. If this rule be observed, it is impossible for the drillings to stick the tools. The drillings can be removed with an ordinary dart-bottom bailer, but it has been found from experience that the patent-bottom bailer is preferable.

CASING A WELL WITH THE FLUID IN THE HOLE.

Casing a gas well with the fluid in the hole can be accomplished in a few hours without the slightest risk to the workmen. On the other hand, several days have been required to case wells that were blowing and on account of the danger from $7 to $10 a day each had to be paid to men to work in the gas.

Should it become necessary to carry casing while drilling, or, in other words, to allow casing to be put in as drilling proceeds, the mud-laden fluid will be of great assistance. The pressure of the fluid on the walls of the well prevents them from caving and freezing the pipe. Underreaming can be accomplished in the same manner. It is sometimes possible with this method to carry the casing from 1,000 to 1,200 feet through a caving, sandy formation, in which a well could not be drilled by other methods.

CASING A "COMBINATION" GAS AND OIL WELL.

In drilling a "combination" gas and oil well by the mud-laden fluid method, the fluid is put into the well just before the gas sand is reached, after which drilling proceeds to a point below the gas sand and the next string of casing is inserted. Before this inner string of casing is seated on the bottom (which can be done either with a packer or shoe, as the case may be), the fluid inside the casing is bailed down, allowing that on the outside of the casing to recede at the same time. A Braden head is attached to the next outer string of casing and packed. The gland of the Braden head is prevented from taking a friction hold on the pipe by two or three small blocks of wood; then when the fluid has been removed to such a depth that its hydrostatic head is less than the gas pressure, the remaining fluid can blow out of the well. The casing is then seated on the bottom and the Braden head bolts, already put in place, are tightened. The seating of the casing in this manner will turn the gas up the outside of the inner casing and expel through the Braden head that part of the fluid between the two casings, so that when the well is cleaned, which will not take more than a few minutes, the valves of the Braden head can be closed and drilling can proceed into the oil below in the usual manner. The arrangement of the casing and Braden head is shown in figure 6.

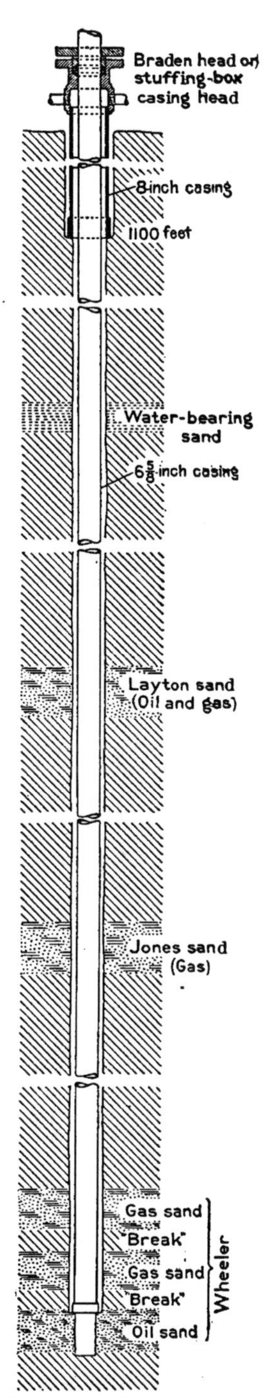

FIGURE 7.—Generalized section of well, showing open passage about casing.

WASTE DUE TO IMPROPER CASING.

To place a casing properly, the drill hole must be large enough to allow the couplings to slip freely down the hole. There is therefore a space of an inch or more between the casing and the walls of the hole. This makes a free path around the casing, which allows water, oil, or gas to pass from one formation to another. (See fig. 7.) The water may drown out the oil or gas, the gas may escape into porous strata, reducing the pressure below commercial value, and the fresh water in any formation penetrated may be spoiled by salt water.

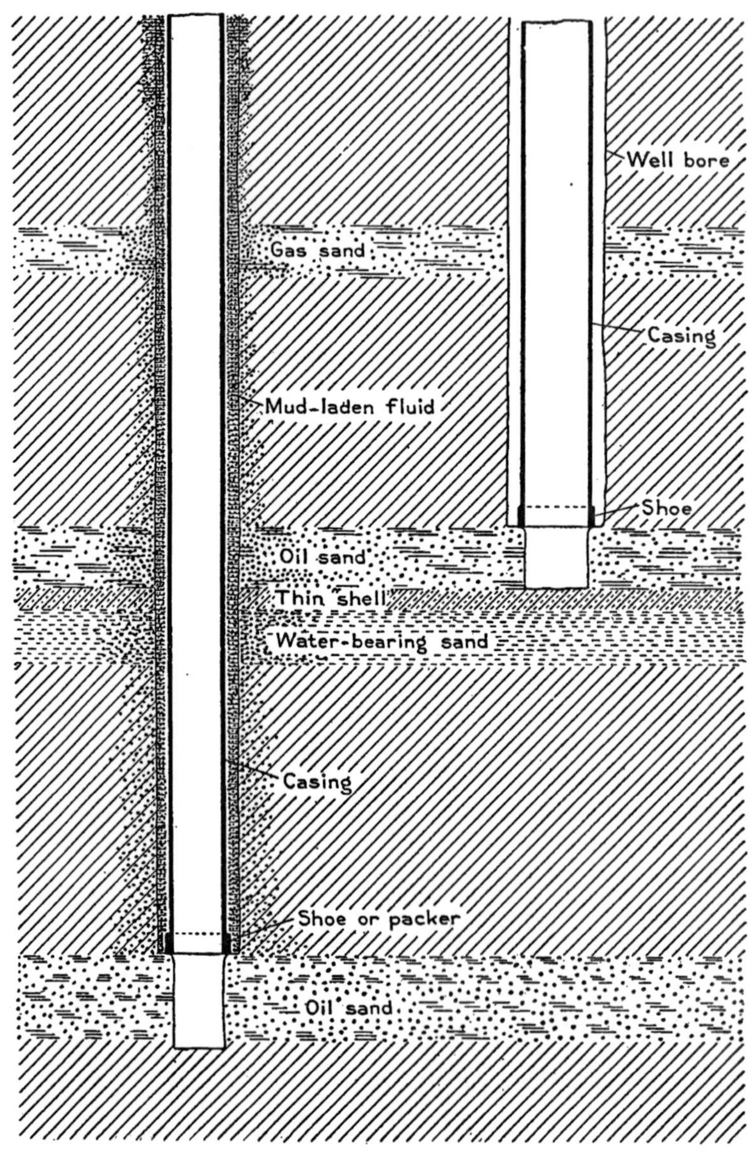

FIGURE 8.—Sealing upper formations by mud-laden fluid method. Well sections are generalized.

Such conditions can not obtain when the mud-laden fluid process is used. The space between the casing and the wall of the hole is filled with the fluid, and all porous formations are sealed with clay, so that oil, water, or gas can not flow from one formation to another, as is shown in figure 6.

PROTECTING UPPER GAS AND OIL FORMATIONS.

It often becomes necessary to case off oil-bearing sands of smaller commercial value in order to reach more valuable sands below, and it sometimes happens that water-bearing beds are in close proximity to the upper oil sands. In the dry-hole method of drilling an extra string of casing is required to properly protect each of these several formations, and as the extra casing delays the work and increases the cost it is seldom put in. When the mud-laden fluid is used extra casing is not necessary, as the muddy liquid entirely fills the space between the casing and the bore of the well (fig. 8). It prevents the waters from going down behind the casing and entering any of the oil sands because the casing must be tightly seated to hold the fluid behind it. In some fields little or no attention is paid to the seating of casing, and as a result water accumulates and rises behind the casing until its head is sufficient to force it under the seat and into the oil-bearing formations. With the mud-laden fluid behind the casing it is impossible for any water to enter the bore hole from the sealed beds and it is much easier for the seat of the casing to retain the mud-laden fluid than to retain clear water. (See fig. 8.)

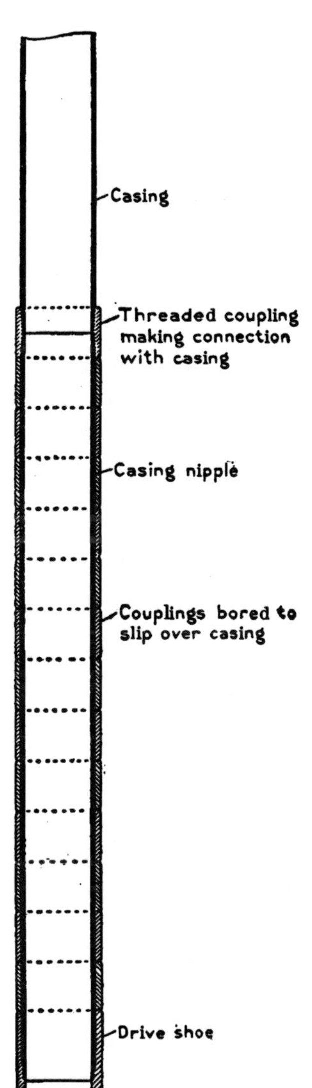

FIGURE 9.—Long casing shoe.

SEALING OFF WATER-BEARING BEDS.

In some fields, in order to exclude water from the oil-bearing formations, cementing is resorted to, as described in another publication[a] of the Bureau of Mines. In some wells as much as 5 to 10 tons of cement are forced in behind the casing in order to shut out the water. The method has come into such general use that many a well is cemented when

[a] Arnold, Ralph, and Garfias, V. R., The cementing process of excluding water from oil wells, as practiced in California: Bull. 32, 1913, 12 pp.

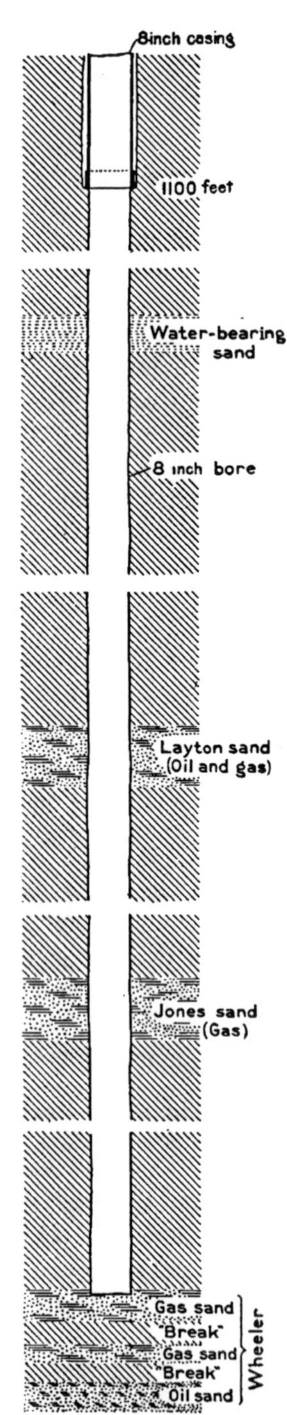

FIGURE 10.—Generalized section of uncased hole permitting communication between formations.

there is really no need of doing so, but because of the cost of drilling the well the operator concludes that the cost of cementing is good insurance in protecting the oil formations. In many cases cementing and trouble from water entering the oil sands can be avoided by using a long shoe, one 6 to 10 feet in length, as conditions may require. This shoe is made by taking a "pup" joint or short piece of casing of the same size as that about to be set, screwing on one end an ordinary casing shoe, and then slipping on reamed-out couplings for that size of casing, until the entire piece of pipe is covered with these collars. A top coupling is then screwed down so that the collars are all clamped between the top coupling and the shoe. By reducing the last 6 or 8 feet of the bore hole to the size of the outside diameter of the collars (fig. 9), contact is obtained for the entire length of this shoe with the bore of the well.

PULLING A LONG CASING SHOE.

If a string of casing has been set with a shoe such as above described, the shoe may cause some trouble in pulling the pipe in the ordinary way when the well is to be abandoned. However, as the bottom joint is reinforced with the collars the string of casing may be started by using a casing spear, which could not possibly be used on pipe without collars, the naked pipe being flattened or expanded by the slips of the spear so that it could not be pulled through the next outer string of casing.

PROTECTING GAS STRATA.

In many parts of the Mid-Continent field the manner in which wells are cased shows that little or no attention is paid to the protection of gas sand.

At some wells the gas is taken from the sands through hundreds of feet of open hole, the last or inner string of casing being stopped far above the gas sand, leaving the lower part of the hole uncased (fig. 10). When such a well is shut in much of the gas escapes into the porous beds below the casing and is wasted. The gas may travel through the porous beds for miles and cause blow-outs in other wells being drilled in the same vicinity, because the drillers have not been expecting to find gas in the sands. Some of these unexpected flows of gas have been ignited, with the result that workmen were seriously injured and the rig and machinery burned.

Another bad practice in casing gas wells is shown in figure 7. The

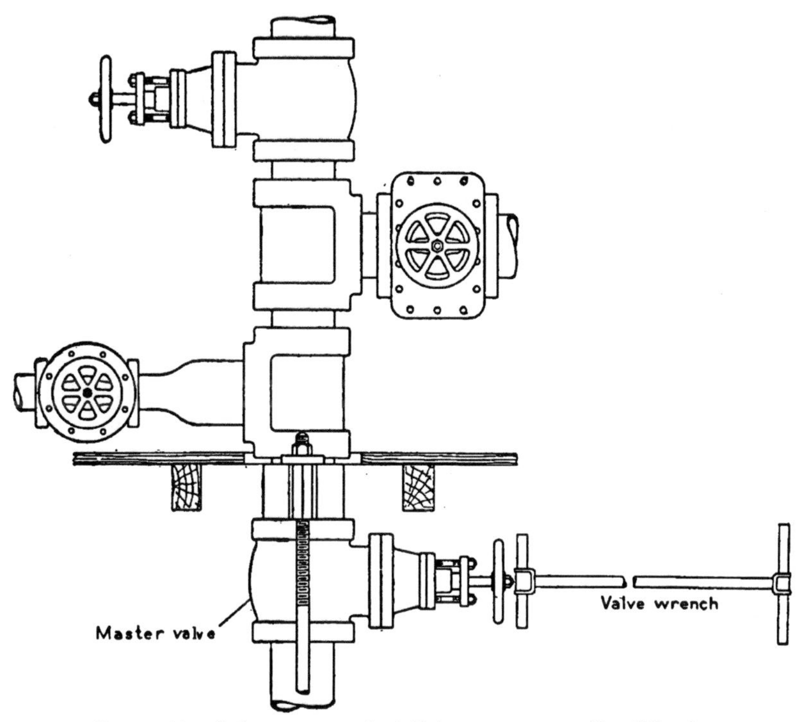

FIGURE 11.—Safe arrangement of fittings on a gas well. Side view.

gas is allowed to flow up between the outside of the inner string of casing and the unprotected walls of the bore hole, and then through the Braden head at the top of the well. The practice should not be permitted, because the gas may escape into any sands not cased.

The proper method of taking gas through a Braden head is to have the string of casing on which the Braden head is placed seated directly above the gas stratum, so that the gas can not reach the formations above the gas sand.

Closely related to the subject of the proper method of casing and finishing a gas well is that of arranging fittings on the head of gas and oil wells.

FINISHING A WELL.

In the ordinary method of shutting in a high-pressure gas well the arrangement of the fittings is such that should the well catch fire from any small leakage around the casing head, or should one of the valves become foul, the flow of gas could not be controlled. When a well equipped in this way catches fire, as may happen, the head of the well has to be broken off by a shot from a cannon before the fire can be extinguished and the well controlled, and the expense or loss to the operator may amount to thousands of dollars.

For a comparatively slight extra first cost a master gate can be placed on the head of the well, as shown in figure 11. This master gate should be packed with asbestos packing throughout, so that if the well catches fire the packing will not be rendered useless. It is also recommended that a pipe T wrench 15 or 20 feet long be attached to the wheel of the master gate in a lateral position, so that the valve can be operated from a point 20 feet from the well. This valve should always remain open, except in case of a fire or an accident to the fittings above the master gate, when the well can be controlled by closing the valve. This comment applies with equal force to the present method of finishing a high-pressure oil well.

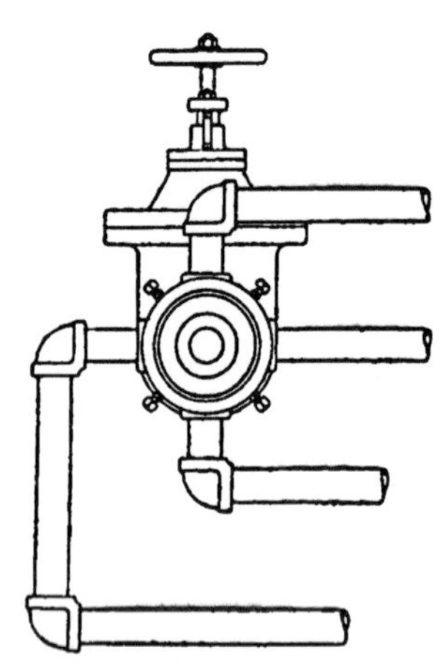

FIGURE 12.—Flow pipes fitted to casing head, with master valve. Top view.

It is the custom in the Mid-Continent field to equip high-pressure oil wells with a casing head, to which are connected flow lines of various sizes. At many wells as many as four 2-inch flow pipes are connected to one casing head, the cap being held in position simply by four set screws. This arrangement would be shown in figure 12 were the master valve omitted.

Should a well catch fire, such an arrangement is entirely inadequate, and in order to get control of the well thousands of dollars may be spent. If, as shown in figure 12, a master valve be placed below the fittings in the manner recommended for a high-pressure gas well, the well can readily be controlled, making it an inexpensive job to put out the fire and repair the damage. It is recommended that a pipe T wrench be attached to the master valve, as illustrated in figure 11.

PUBLICATIONS ON PETROLEUM TECHNOLOGY.

The following Bureau of Mines publications may be obtained free by applying to the Director, Bureau of Mines, Washington, D. C.:

BULLETIN 19. Physical and chemical properties of the petroleums of the San Joaquin Valley, Cal., by I. C. Allen and W. A. Jacobs, with a chapter on analyses of natural gas from the southern California oil fields, by G. A. Burrell. 1911. 60 pp., 2 pls., 10 figs.

TECHNICAL PAPER 3. Specifications for the purchase of fuel oil for the Government, with directions for sampling oil and natural gas, by I. C. Allen. 1911. 13 pp.

TECHNICAL PAPER 10. Liquefied products from natural gas; their properties and uses, by I. C. Allen and G. A. Burrell. 1912. 23 pp.

TECHNICAL PAPER 25. Methods for the determination of water in petroleum and its products, by I. C. Allen and W. A. Jacobs. 1912. 13 pp., 2 figs.

TECHNICAL PAPER 26. Methods of determining the sulphur content of fuels, especially petroleum products, by I. C. Allen and I. W. Robertson. 1912. 13 pp., 1 fig.

TECHNICAL PAPER 32. The cementing process of excluding water from oil wells, as practiced in California, by Ralph Arnold and V. R. Garfias. 1913. 12 pp., 1 fig.

TECHNICAL PAPER 36. The preparation of specifications for petroleum products, by I. C. Allen. 1913. 12 pp.

TECHNICAL PAPER 37. Heavy oil as fuel for internal-combustion engines, by I. C. Allen. 1913. 36 pp.

TECHNICAL PAPER 38. Wastes in the production and utilization of natural gas, and means for their prevention, by Ralph Arnold and F. G. Clapp. 1913. 29 pp.

TECHNICAL PAPER 42. The prevention of waste of oil and gas from flowing wells in California, with a discussion of special methods used by J. A. Pollard, by Ralph Arnold and V. R. Garfias. 1913. 15 pp., 2 pls., 4 figs.

TECHNICAL PAPER 49. The flash point of oils; methods and apparatus for its determination, by I. C. Allen and A. S. Crossfield. 1913. 31 pp., 2 figs.

TECHNICAL PAPER 51. Possible causes of the decline of oil wells and suggested methods of prolonging yield, by L. G. Huntley. 1913. 32 pp., 9 figs.

TECHNICAL PAPER 57. A preliminary report on the utilization of petroleum and natural gas in Wyoming, by W. R. Calvert, with a discussion of the suitability of natural gas for making gasoline, by G. A. Burrell. 1913. 23 pp.

O

CPSIA information can be obtained
at www.ICGtesting.com
Printed in the USA
LVOW10s1534220318
570807LV00035B/690/P

9 781332 819508